WEIRD SPACE SCIENCE

INTRODUCTION TO SPACE

THE BREAKDOWN

VIRGINIA LOH-HAGAN

45TH PARALLEL PRESS

Published in the United States of America by Cherry Lake Publishing Group
Ann Arbor, Michigan
www.cherrylakepublishing.com

Photo Credits: © Delpixel/Shutterstock, cover, 1; © Nikolaeva, cover, 1; © NASA/JPL-Caltech/MSSS, cover, 1; © sripfoto/Shutterstock, cover, 1, 6; © Vadim Sadovski/Shutterstock, 8-9; © PlanilAstro/Shutterstock, 11; © Alhovik/Shutterstock, 13; © Noel Powell/Shutterstock, 14; © Dima Zel/Shutterstock, 17, 20; © achinthamb/Shutterstock, 18-19

Graphic Element Credits: Cover, multiple interior pages: © WindAwake/Shutterstock, © kichikimi/Shutterstock, © Your Local Llamacorn/Shutterstock, © Marishiav/Shutterstock, © Supza/Shutterstock, © HSSstudio/Shutterstock, © Olive Kitt/Shutterstock, © KKarance/Shutterstock, © Theo_hrm/Shutterstock, © Rodin Anton/Shutterstock, © Petr Vaclavek/Shutterstock, © Nikolaeva/Shutterstock, © Evgenii Doljenkov/Shutterstock

45th Parallel Press is an imprint of Cherry Lake Publishing Group.

Library of Congress Cataloging-in-Publication Data has been filed and is available at catalog.loc.gov

Cherry Lake Publishing Group would like to acknowledge the work of the Partnership for 21st Century Learning, a Network of Battelle for Kids. Please visite Battelle for Kids online for more information.

Printed in the United States of America

Dr. Virginia Loh-Hagan is an author, university professor, and former classroom teacher. She's currently the Director of the Asian Pacific Islander Desi American Resource Center at San Diego State University. She loves science-fiction stories. She lives in San Diego with her very tall husband and very naughty dogs.

TABLE OF CONTENTS

BREAK IT DOWN

1. DECODE IT

- Look at each letter.
- Look at letter patterns.
- Blend sounds.
- Try other possible sounds.

2. BUILD VOCABULARY

- Listen to the word.
- Match the sounds to a word you know.
- Look for context clues for words you don't know.
- Look up the meaning in a glossary or dictionary.

3. MAKE CONNECTIONS

- Name other examples.
- Draw the idea.
- Compare ideas.
- Ask questions.

TEST IT OUT

WORDS YOU'LL LEARN

Look over the list of vocabulary words. Are there any you already know?

astronauts
galaxies
intergalactic medium (IGM)
NASA
orbit
solar systems
universe

GET STARTED

Word parts can be clues for how to read a word. They can be clues to meaning, too.

- Look at the letters.
- Look for familiar word parts.
- Chunk long words by word part.

The word *astronauts* has 2 word parts: *astro-* and *-nauts*. It is pronounced: AA-struh-nawts.

- *Astro-* is from the Greek aster. It means star.
- *-nauts* is from the Greek naútēs. It means sailor.

All kinds of weird science happens in space.

Outer space is the area beyond Earth's sky.

The **universe** is huge.

It's everything that exists.

It includes planets and stars.

It includes everything on Earth.

It includes billions of **galaxies**.

BUILD VOCABULARY

Galaxies (GAA-luhk-seez) are big groups of stars.

- They are in the universe.
- Our sun is in the Milky Way galaxy.
- Other forms: galaxy, galactic, intergalactic

Galaxies are huge space collections.

They are made up of billions of stars, gas, and dust.

Galaxies include **solar systems**.

Solar means having to do with a sun.

A system is a set of things
that work together.

A solar system is a sun and the objects that **orbit** it.

This is the curved path of an object around a star, planet, or moon.

There's a lot of space between stars and galaxies. It is mostly empty.

Some gas, dust, light, heat, and rays are there, too. This is **intergalactic medium (IGM)**. Medium means something in the middle.

DECODE IT

The word *intergalactic* has 3 word parts.

- Break it into chunks: inter-galact-ic.
- *inter-* means between or among.
- *-ic* means of or related to.
- *-galact-* is a form of galaxy.
- *Intergalactic* is pronounced in-tuhr-guh-LAK-tik.

Scientists study space.

They want to see if life is possible on other planets.

MAKE CONNECTIONS

Scientists define "life" in different ways. Some think a thing must breathe to be alive. Some think it must grow.

- Think about things that are alive.
- Is fire alive? Why or why not?
- Do you think there is life beyond Earth?

They send up astronauts to explore space.

Astronauts are people trained to fly to space.

NASA is the U.S. space program.

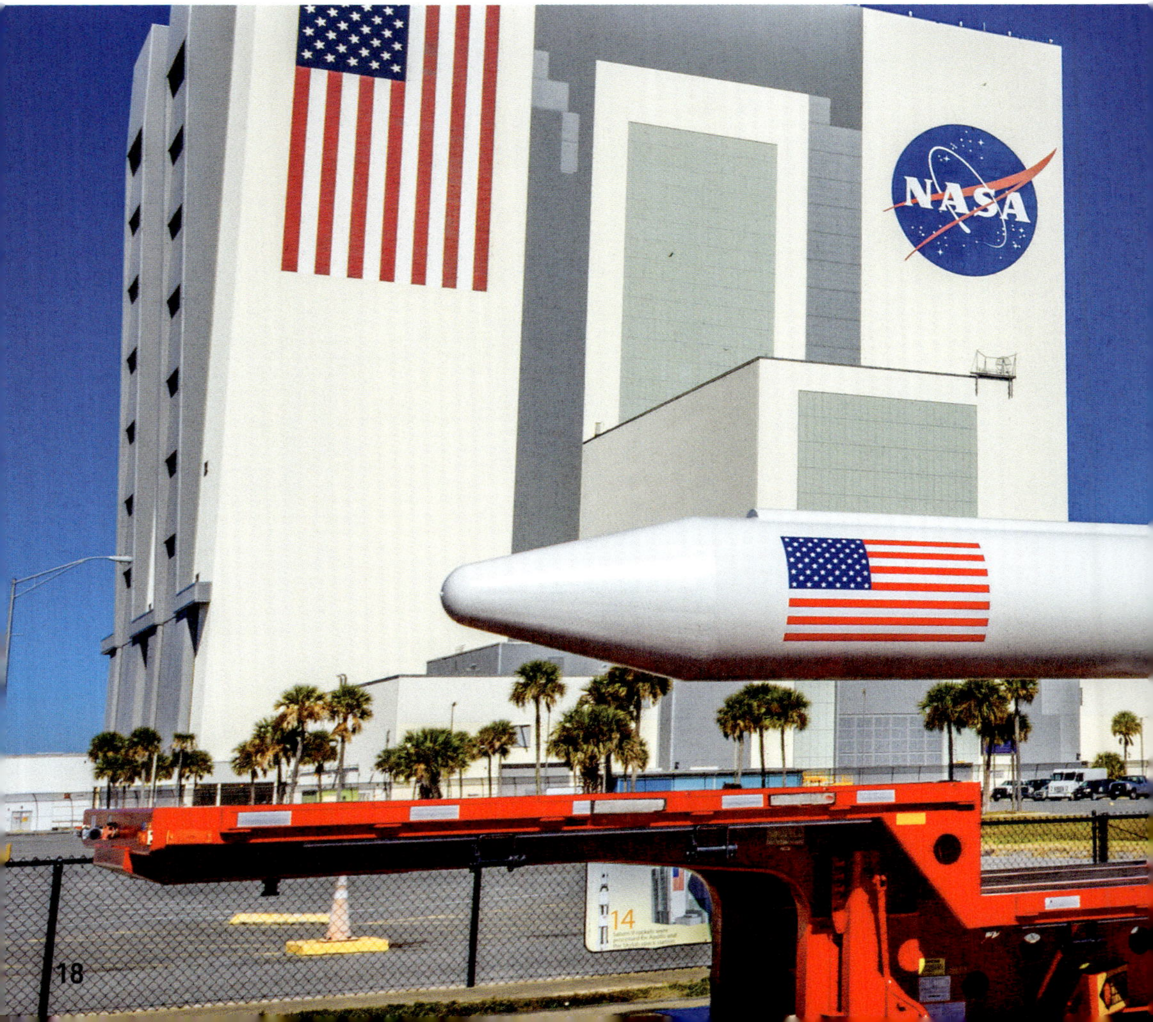

NASA stands for National Aeronautics and Space Administration.

We know a lot about space.

But there's so much left to learn!

PUT IT TOGETHER

Detail:
Astronauts travel to space.

Detail:
Scientists study space. NASA is the U.S. space program.

Main Idea:
There is a lot to learn about space.

Detail:
The universe is everything that exists. It includes planets and stars. It includes galaxies.

Detail:
Space between stars and galaxies is filled with dust, light, heat, and rays.

Detail:
Galaxies are made up of stars, gas, and dust. They hold solar systems.

LEARN MORE

BOOKS

Loh-Hagan, Virginia. *Introduction to Space*. Ann Arbor, MI: 45th Parallel Press, 2025.

Loh-Hagan, Virginia. *Weird Science: Space*. Ann Arbor, MI: 45th Parallel Press, 2022.

Tyson, Neil deGrasse. *StarTalk: Young Readers Edition*. Washington, DC: National Geographic Kids, 2018

ONLINE

Search these online sources with an adult:

- NASA | Games and Interactives
- NASA+ | Explore Topics

GLOSSARY

astronauts (AA-struh-nawts) people whose job is to travel into space

galaxies (GAL-uhk-seez) huge space collections made of billions of stars, gas, and dust

intergalactic medium (IGM) (in-tuhr-guh-LAK-tik MEE-dee-uhm) gas, dust, light, heat, and rays in the space between galaxies

NASA (NAA-suh) the federal U.S. space program

orbit (OR-buht) curved path of an object around a star, planet, or moon

solar systems (SOH-luhr SIS-tuhmz) suns and the objects that orbit them

universe (YOO-nuh-vuhrs) all things that exist

INDEX